YOUR KNOWLEDGE HAS VALUE

- We will publish your bachelor's and
  master's thesis, essays and papers

- Your own eBook and book -
  sold worldwide in all relevant shops

- Earn money with each sale

Upload your text at www.GRIN.com
and publish for free

Amor James

Human Contribution to Rising Global Temperatures and Rising Sea Levels

GRIN Verlag

Bibliografische Information der Deutschen Nationalbibliothek:

Die Deutsche Bibliothek verzeichnet diese Publikation in der Deutschen National-
bibliografie; detaillierte bibliografische Daten sind im Internet über http://dnb.d-
nb.de/ abrufbar.

Imprint:

Copyright © 2011 GRIN Verlag GmbH
Druck und Bindung: Books on Demand GmbH, Norderstedt Germany
ISBN: 978-3-656-59225-9

This book at GRIN:

http://www.grin.com/en/e-book/268210/human-contribution-to-rising-global-tem-
peratures-and-rising-sea-levels

Human Contribution to Rising Global Temperatures and Rising Sea Levels

ABSTRACT

The causes of rising sea levels is in a state of continuous debate, but increasing scientific research suggests that human activity is at least partial, if not the predominate cause of the rapid rise in ocean levels worldwide. High concentrations of CO2 and other GHGs released into the atmosphere from human consumption, industrialization and manufacturing attribute to rising global temperatures, which is in turn absorbed by the sea, causing the warming and expansion of oceans. The detrimental effects of warming oceanic temperatures is multifaceted. It leads to ocean freshening as a result of melting glaciers in the Arctic and Antarctic, which raises sea levels and it negatively impact on the survival of humans, flora, fauna and marine ecosystem. Furthermore, higher temperatures in global seas attribute to thermal expansion and eustatic rise, which is a key cause behind why the oceans are taking up more of the Earth's surface. Thermal expansion also causes irreversible degradation of biodiversity and human habitat in coastal areas. Thus, this paper will explore how intensive human activity contributes to climate change which heats up sea temperatures and lead to ocean expansion and rising shore levels around the globe.

EUSTATIC RISE

Over 95 percent of the global scientific community confirms that climate change is largely attributed by increasing emissions of heat trapping GHGs from human activity, which causes eustatic rise in the oceans. In the past 100 years, tide gauge readings and satellite measurements show that the Global Mean Sea Level (GMSL) has gradually risen between 10 to 20 centimeters.[1] Moreover, research indicates that eustatic levels worldwide have risen at an average rate of 3.5 millimeters per year, since the start of the 1990s.[2] This means that the rate of rising ocean levels within the past 20 years is double the average speed of the preceding 80 years. According to an article in *Energy and Environment,* satellite altimetry measurements show that the GMSL has risen approximately 2cm per year between 1992 and 2000. Moreover, the mean rate rose to 3.2 cm per year in 2003, 5 cm per year in 2008 and 6 cm per year at the present day.[3] There are severe consequences of climate change and rising eustatic levels.

One of the key factors in rising sea levels is the melting of ice caps in the Arctic and Antarctic due to warming average global temperatures. Researcher Thorarinsson suggests that after the last *Little Ice Age* (LIA) at the beginning of the 19th century, the Earth's surface temperature rose which led to glaciers to recede and sea levels to rise. Between the mid-1800s and 1900s, a calculated eustatic rise of 0.5 mm per year was a direct correspondent of retreating ice caps.[4] To be more precise, sea levels rose approximately 11 cm between the years of 1850 to 1950, and rapid rise of shore levels were the most evident in Northwestern Europe.[5] Therefore, it

---

[1] "Sea Level Rise." *National Geographic.* 2013. (accessed 3 Aug. 2013).
<http://ocean.nationalgeographic.com/ocean/critical-issues-sea-level-rise/>.
[2] Ibid.
[3] Nils-Axel, Morner. "Sea Level Changes Past Records and Future Expectations." *Energy and Environment* 24. 3&4 2013. (accessed 2 Aug. 2013).
[4] Ibid.
[5] Ibid.

can be speculated that the significant custatic rise between the mid-19th to 20th century directly corresponds to the boom of the industrial and agricultural revolution that depended on intensive burning of fossil fuels. The above data offers proof that human activity plays a role in the rapidly rising global temperatures and rising shores levels.

RISING GHG EMISSION AND HIGHER SEA LEVELS

Throughout the millennia, there have been long periods of globally warmer and cooler climatic conditions on the Earth's surface related to solar flares and other activities of the sun. Within the last 10,000 years, dramatic climatic shifts happened on the planet. When global temperatures rose by just 3 degrees Celsius in 4000 BC, the Sahara region changed from a lush savannah to an arid desert within a few centuries. The *Modern Warm Period* that took place after the LIA during the 1550s to 1850s gave rise to the massive proliferation of the human population and the heavy oil burning era that originated from the industrial revolution.[6] A prevalent geological trend that researchers have connected to changes in the global climate is a high volume of CO2 concentrations in the Earth's atmosphere. A 30% increase of carbon emissions beginning in the 20th century is a direct result of intensive fossil fuels consumption in human activity.[7] According to the Intergovernmental Panel of Climate Change (IPCC), ongoing increase of CO2 concentrations in the atmosphere will lead to more frequent effects of climate change, manifesting in the form of severe winters, long heat waves, eustatic rise, floods and droughts. There are several correlations between increased GHG emissions and rising ocean levels.

Growing atmospheric concentrations of CO2 and other heat trapping gases produced from human activity leads to increased infrared absorption capacity of the troposphere, which

---

[6]   Harry, N.A., Priem. "Climate Change and Carbon Dioxide: Geological Perspective." *Energy and Environment* 24.3 & 4 (2013) (accessed 6 Aug. 2013).
[7]   Ibid.

enhances the greenhouse effect. When the heat of the sun is confine on the surface Earth by GHGs, global temperatures and the oceans heat up. The IPCC observed a rise in the average global surface temperature of 0.8 degrees Celsius since 1950 and the collective foresees an upward climb in global temperatures of 0.2 degrees Celsius per decade in the 21st century.[8] What is more is that global warming also heats up oceans worldwide, because up to 80 percent of the Earth's surface temperature is absorbed by the seas.[9] Consequently, when the temperature of the oceans heat up, the water mass expands. Based on tide gauge records, the IPCC found that the average eustatic rise in the 20th century was between 1.5 and 2.0 cm per year.[10] Moreover, in the ongoing debate of how significantly carbon dioxide causes climate change, the IPCC claims that human-made GHG concentrations, particularly CO2, is the only logical explanation to warming global temperatures and rising sea levels in the second half of the 20th century.[11] Therefore, the above data adds to the body of evidence that show, high-carbon human activities force changes in the climate linked to ocean thermal expansion to happen at a faster pace than the Earth's natural mechanisms.

THERMAL EXPANSION AND OCEAN FRESHENING

Since global oceans absorb approximately 80 percent of the heat trapped on the planet, the world's warming seas are expanding and taking up more of the Earth's surface. General circulation models (GCMs) developed by Jackett et al. suggest that over half of the rising sea levels predicted for the future can be attributed by oceanic thermal expansion due to the enhanced greenhouse effect on Earth.[12] Furthermore, the IPCC and a majority of the global

---

[8] Harry N.A., Priem. "Climate Change and Carbon Dioxide: Geological Perspective".

[9] "Sea Level Rise." *National Geographic.*

[10] Walter, Munk. "Ocean Freshening, Sea Levels Rising," *Science.* 300, 2003. (accessed 5 Aug. 2013).

[11] Harry, N.A., Priem. "Climate Change and Carbon Dioxide: Geological Perspective".

[12] D.R., Jacket et al. "Thermal Expansion in Ocean and Coupled General Circulation Models." *Journal of Climate.* 13.8, 2000 (accessed 3 Aug. 2013).

community believes that rising eustatic levels are largely caused by the concurrent rise in global temperatures over the past century, that directly contribute to marine thermal expansion.[13] Additionally, in 1995, the IPCC suggest that the import of fresh water from different continents as a result of melting sea ice sheets is an additional contributor to eustatic rise.[14] Munk refers to the increase of water mass in the world's oceans as *sea freshening,* partially because melting glaciers add fresh water to marine ecosystems.

A key factor behind rising eustatic levels is ocean thermal expansion from warming sea temperatures. General circulation models are used to observe ocean *heat uptake* in order to demonstrate that temperatures in global seas are rising with the progression of climate change. GCMs are based on different sets of mathematical equations that work to calculate climate controlling processes, and forecast the climate under various CO2 emission scenarios.[15] Jackett et al. calculate the heat uptake of global marine systems by applying the temperatures at the base of the ocean mixed layer, against data on the interior ocean thermal status, to predict the sea surface temperature (SST).[16] Jackett's GCMs offer evidence that show a rise in SST globally. The link between rising sea temperatures and shore levels can be explained by the thermal expansion phenomena; heated temperatures cause water particles in the sea to move at greater speeds and velocity,[17] thus achieving a greater separation rate. Great separation rate cause the world's oceans to further expand across the surface of the planet.

In a study conducted by Domingues et al. the team researched the effects of ocean surface warming since 1880. Their results are found using measurements of thermosteric sea level shifts

---

[13]   Walter, Munk. "Ocean Freshening, Sea Levels Rising".
[14]   Walter, Munk. "Ocean Freshening, Sea Levels Rising".
[15]   Harry, N.A., Priem. "Climate Change and Carbon Dioxide: Geological Perspective".
[16]   D.R., Jacket et al. "Thermal Expansion in Ocean and Coupled General Circulation Models".
[17]   Ibid.

in the upper ocean layer, above 300 metres in depth and in the internal sea layers, 700 m below the sea surface.[18] This research shows that the largest sample of thermostatic sea level change happens in the upper 300 m, and thermal expansion in the deep seas occur much slower, at a rate of approximately 0.05 mm per year.[19] This is because oceanic conditions at the upper sea level is changed by varying seasonal climatological boundary conditions of temperature, salinity, and wind stress.[20] Further, earlier effects of global warming may still be spreading to the marine interior. Using similar methods, Marcelja's research team calculated a global average rate of ocean thermal expansion with the advection-diffusion model. They found that between 1880 till now, the Earth has had a 48mm thermal expansion rate.[21] More of the thermostatic expansion occurred within the past 50 years.

Finally, evidence to proof that the rapid receding rate of ice caps in the Arctic and Antarctic results from the global climate crisis is found in salinity measurements of the oceans. Antonov et al. found that the mean salinity of the sea worldwide has decreased between 1954 and 1997.[22] The combination of higher sea levels, increased oceanic temperatures and decreased salinity suggest that melting sea ice sheets and continental fresh water deposits is attributing to global eustatic rise. The estimated total sea ice volume is $30,000 km^3$ and Johannessen et al. show a calculation of sea ice recession at $150 km^3$ per year, within the past 20 years.[23] This rate of annual glacial loss adds over $135 km^3$ of fresh water into global oceans each year, which result in

---

[18] S. Marcelja. "The Timescale and Extent of Thermal Expansion of the Oceans Due to Climate Change," *Ocean Science Discussions*. 6.3: 2009 (accessed 6 Aug. 2013). <http://www.ocean-sci-discuss.net/6/2975/2009/osd-6-2975-2009.pdf>.

[19] Ibid.

[20] D.R., Jacket et al. "Thermal Expansion in Ocean and Coupled General Circulation Models".

[21] Ibid.

[22] Walter, Munk. "Ocean Freshening, Sea Levels Rising".

[23] Ibid.

eustatic rise levels of 1.9mm per year.[24] Ocean freshening due to global warming is happening so quickly that the North Pole has melted to form a lake as of July 2013, the North Pole Environmental Observatory (NPEO) claims. The NPEO explains that July marks the warmest month of the year in the Arctic. And this year, temperatures were 1 to 3 degrees Celsius higher than average, which has created a lake of meltwater sitting on top of the already thinning ice sheet.[25] Arctic habitat, wildlife and native communities are direct victims of shrinking glaziers and ocean thermal expansion due to climate change.

CONCLUSION

Different studies and measurement methods offer variations in the scientific data of the effects of high GHG emissions levels in relation to eustatic levels, ocean salinity levels and heat-uptake levels in the global oceans. However, it cannot be denied that the global scientific community almost unanimously agrees that intensive human activity contribute to climate change and rising sea levels. In fact, there are not many reputable arguments that can dispute the fact that human activity revolved around the burning fossil fuels releases high concentrations of emissions that trap heat to speed up effects of climate change. Evidence show that average global temperature is rising at a much faster pace than the Earth's natural climatic cycles. Moreover, the world's oceans are absorbing 80% of the heat trapped on the Earth's surface. When oceanic temperatures heat up, eustatic levels rise at a faster rate than the global scientific community has previously anticipated. Finally, the warming and expanding oceans threaten the livelihoods of natural marine ecosystems and coastal communities worldwide. Consequences of rising sea levels include species extinction, loss of natural habitat and destruction of human developments,

---

[24] Ibid.

[25] Nick, Visser. "North Pole Melting Leaves Small Lake at The Top Of The World." *Huffington Post Green*. 25 Jul 2013. (accessed 5 Aug. 2013). <http://www.huffingtonpost.com/2013/07/25/north-pole-melting-leaves_n_3652373.html>.

and these affects are not irreversible. However, the course of human activity that attributes to the global climate crisis can be drastically changed in order to slow the rise of global temperatures and thermal expansion of the world's oceans.

Citations

Degener, Richard. "Sea's Rising Acid Levels Threaten Shellfish - Lautenberg Bill Would \provide $3 Million for Research." *Press of Atlantic City*. (2007): n. page. Web. 1 Aug. 2013. <http://infoweb.newsbank.com.ezproxy.library.ubc.ca/iw-search/we/InfoWeb?p_product=AWNB&p_theme=aggregated5&p_action=doc&p_docid=11CB5589089EDD08&d_place=PACB&f_issue=2007-11-03&f_publisher=>.

Jacket, D.R., T.J. McDougall, M.H. England, and A.C. Hirst. "Thermal Expansion in Ocean and Coupled General Circulation Models." *Journal of Climate*. 13.8 (2000): 1384-1405. Web. 3 Aug. 2013.

Marcelja , S. "The Timescale and Extent of Thermal Expansion of the Oceans Due to Climate Change." *Ocean Science Discussions*. 6.3 (2009): 2975-2992. Web. 6 Aug. 2013. <http://www.ocean-sci-discuss.net/6/2975/2009/osd-6-2975-2009.pdf>.

Morner, Nils-Axel. "Sea Level Changes Past Records and Future Expectations." *Energy and Environment*. 24.3&4 (2013): 509-536. Web. 2 Aug. 2013.

Munk, Walter. "Ocean Freshening, Sea Levels Rising." *Science*. 300. (2003): 2041-2043. Web. 5 Aug. 2013.

Priem , Harry N.A. "Climate Change and Carbon Dioxide: Geological Perspective." *Energy and Environment*. 24.3 & 4 (2013): 361-380. Web. 6 Aug. 2013.

"Sea Level Rise." *National Geographic*. 2013: n. page. Web. 3 Aug. 2013. <http://ocean.nationalgeographic.com/ocean/critical-issues-sea-level-rise/>.

Visser, Nick. "North Pole Melting Leaves Small Lake at The Top Of The World." *Huffington Post Green*. 25 Jul 2013: n. page. Web. 6 Aug. 2013. <http://www.huffingtonpost.com/2013/07/25/north-pole-melting-leaves_n_3652373.html>.